AF772157

Omar Selt

# Methods for combinatorial optimization -With examples

Omar Selt

# Methods for combinatorial optimization -With examples

## Methods for combinatorial optimization

Noor Publishing

**Imprint**



Cover image: www.ingimage.com

Publisher:
Noor Publishing
is a trademark of
Dodo Books Indian Ocean Ltd. and OmniScriptum S.R.L publishing group

120 High Road, East Finchley, London, N2 9ED, United Kingdom
Str. Armeneasca 28/1, office 1, Chisinau MD-2012, Republic of Moldova, Europe
Managing Directors: Ieva Konstantinova, Victoria Ursu
info@omniscriptum.com

Printed at: see last page
**ISBN: 978-620-0-04637-6**

# Methods for combinatorial optimization

**With examples**

**Pr.Dr.Omar Selt**

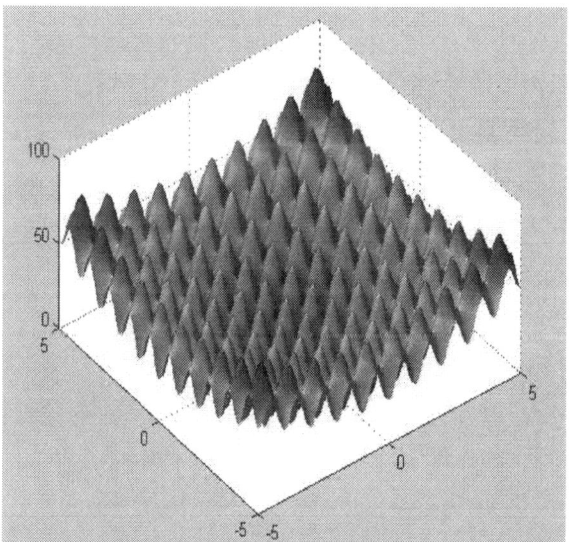

## Introduction

A combinatorial optimization problem consists of finding the minimum s of a function f (economic functions), with real or integer variable on a finite set S, the elements of S verifying certain constraints are called feasible solutions; among which, the optimal solution is included.

Optimization is choosing an efficient method capable of producing an optimal solution in a reasonable computational time. The different resolution methods developed can be classified into two categories: exact methods which guarantee the completeness of the resolution and approximate methods which lose completeness to gain efficiency.

## Optimization problems

Complexity theory is built on decision problems. Optimization problems involve optimizing a measure over a set of feasible solutions. An optimization problem Π will be defined by the following data:

A set I of instances.

A function **S** such that for all **x** ∈**I**, **S(x)** represents the set of feasible solutions for **x**

An integer-valued function **m** is defined on all pairs **x** ∈**I** and **y** ∈**S(x)**.

This measure **m** represents the objective function of a solution **y** instance **x**.

min, max }objective ∈{specifying whether Π is a minimization or maximization problem.

To any instance **x**, we associate a solution **y*(x)** and the value **m*(x)** defining an optimum. If Π is a minimization problem,**m*(x)=min{ m(x, y) |y∈S(x) }**and **y*(x)** is a solution verifying **m(x, y*(x))** = **m*(x)**.

To such an optimization problem Π, we can associate 2 different versions depending on the nature of the expected result (function calculated by a resolution algorithm):

$\Pi_C$ ( default Π), the construction problem. For each instance **x**, we must provide as output the value **m*(x)** and a solution **y*(x)** achieving this optimum.

- $\Pi_D$, the associated decision problem. An instance of $\Pi_D$ is an instance **x** of Πmore integer **K**. The question is: does there exist a solution **y** verifying **m(x, y) ≤K** for the problem of **min ( resp . m(x, y) ≥K for max)**

Problems are not decision problems, and therefore cannot belong to NP, let alone to NP-complete. The notion of reduction between these problems is necessary to intuitively define that a problem Π is at least as difficult as a problem that reduces to Π.

A problem Π'is reduced to whether we can write a procedure to solve knowing a sub-procedure to solve Π. We can then define NP-difficulty as follows.

**Decision problem**

It is a problem whose results can only take one of two values: TRUE or FALSE.

**The NP class** ✓

The class of NP problems (NP for Polynomial Nondeterminacy) is the class of decision problems that can be solved by a nondeterministic Turing machine in polynomial time.

Among the NP problem class, we distinguish two large classes:

The class of polynomial problems (the P class) and the class of NP-complete problems (the NPC class).

**Class P:** A problem is said to belong to class P ✓ if there exists a polynomial algorithm to solve it. Problems in class P are said to be easy.

**The NP-Complete class:** ✓

A decision problem is said to be NP-complete if every problem in the class NP can be polynomially resolved to it.

Not all NP-hard problems are of the same - difficulty. For example, we encounter problems that cannot be solved in polynomial time with a binary coding, but which can be solved with a unary coding; these problems are called NP-hard in the ordinary sense or simply NP-hard. Algorithms for this class of problems are called pseudo-polynomial. For other problems, we may not find a polynomial algorithm, whatever the coding; these are called NP-hard in the strong sense.

**Definition**

A problem $\Pi$ is said to be NP-hard or NP-hard if only for any problem $\Pi_0 \propto \Pi$.

For an optimization problem $\Pi$. We clearly have $\Pi_D \propto \Pi_C$. For this reason, we use in practice the following property.

**Proposition**

Let the $\Pi$an optimization problem be given.

If its associated decision problem $\Pi_{D\ is} \in$NP-complete, then $\Pi \in$NP-hard.

Most scheduling problems are NP-hard because they are optimization problems. However, there are some special cases for which polynomial algorithms have been found, such as for the problem:

In the field of combinatorial optimization, belonging to the class of NP-hard problems does not at all mean the end of the study of a problem.

**The exact methods**

The resolution comes down to exploring all possible sequences through a search tree. The computational time required for such a method generally increases exponentially with the size of the problem to be solved. They come from operations research, the goal of these methods is to find the optimal solution to the problem in the shortest possible computational time, so the resolution comes down to exploring all possible sequences through a search tree.

To improve the efficiency of the search, techniques are used to calculate bounds that allow the early pruning of branches that lead to failure, and heuristics are used to guide the choice of variables and values during the exploration of the tree. Among the exact methods, we distinguish:

**Complete enumeration**

We seek an optimal schedule across all possible schedules, and full enumeration generates the possible schedules. In the case of non-preemptive scheduling problems on a machine, there are **n! Possible task execution** sequences, so for the corresponding **m-**machine problem there are **(n!)** $^{m}$ **possible execution** sequences, this number is very large. Although in practice, due to the constraints of the problem, many schedules may be inadmissible, and elimination procedures may sometimes be used to see if the non-optimality of one schedule implies the non-optimality of another, not yet constrained, the solution time can be prohibitively long even when using the most powerful and fastest computers. It is therefore clear that seeking an optimal solution using full enumeration is not suitable even for small problems. This is why it was necessary to resort to intelligent methods.

**Combinatorial Analysis**

This method examines the effect of a minor change in a particular sequence on the value of that sequence). Combinatorial analysis can lead to very efficient algorithms that produce an optimal ordering in a predictable number of steps where the latter grows at most polynomially as a function of the problem size. This method examines the effect of a minor change to a particular sequence on the value of that sequence.

**Dynamic programming**

It is a method based on a recursive approach, that applies to any problem that can be divided into a sequence of problems, the solution of a problem of order **n** is simply expressed in terms of solutions of problems of order **( n – 1)**.

The dynamic programming method is a generic method for optimization problems and in particular it has been proven to be a flexible and powerful tool for dealing with scheduling problems, it was first applied to scheduling problems by **CARP**.

This method consists of breaking down the problem posed into sub-problems (or phases), then establishing a recursive equation expressing the solution or the sub-problem of order **K** as a function of that of the sub-problem of order **(K- 1)**. Thus, the last sub-problem would be of order **(n)**. In this case, it is the problem to be solved in the example below we illustrate the dynamic programming approach.

**1// f** $_i$, or we minimize the total cost $\sum' f(C_J)$

This recursive function was suggested by **HEL** and **KARP** and then **[Law 64]**

let **J** $\subseteq$**I** = {**1, 2, 3..........,n be** }a set of tasks and **f *(J)** the minimum total cost of scheduling **J's tasks** over the period $\left[0, \sum_{i \in j} P_i\right]$,

the objective is to find **f*(J)** by solving the recursive equations.

$$f^*(J) = \min_{i \in j} \{f^*(J - \{i\}) + f_i((\sum_{i \in J}' P_i)\}.$$

which are initialized by **f*( ∅) = 0**

Dynamic programming is an implicit enumeration method, better than full enumeration, but whose memory requirements are often very large.

**Method by separation and evaluation**

These techniques were first applied to scheduling problems by LominickiI and Ignal. This method represents the possible solutions by a "search tree" that must be explored from top to bottom, using a lower bound ( LowerBound ), the optimal solution, determined beforehand, we evolve at each node more than the lower bound and only explore those which improve this lower bound, this method, therefore, makes it possible to reduce the number of cases to explore. This number depends essentially on the quality of the lower bound adopted at the start.

**The search tre**

The search tree is the schematic representation allowing us to systematically and explicitly list all the **n!** Permutations of **n** tasks. It appeared as a set of nodes arranged in hierarchical layers and numbered respectively from 0 to (n-1) from the highest to the lowest level.

Layer 0 contains a single node, called the root node, which we denote by **S** . Layer **n-1** is composed of **n!** Nodes are called leaf nodes, and the nodes of the other layers are called internal nodes.

Each node is a scheduling set. In particular, **S** is the set of all schedulings ( Card (S) = n!), and any leaf node is a singleton we denote by $I_y$ the set of tasks already assigned in node **Y** ($I_S = \phi$ and $I_F = I$ if **F** is a leaf node) by 0 a node of layer **i**, and by **D** a node of layer **i+1** where **i ≠n-1**

**D** is said to be a successor (0) if it is directly derived from it, i.e. $I_D - I_0 = \{t\}$, where (t) is the branching task from node 0 to node **D,** and the tasks are arranged identically in **D** as in 0.

Every node has **(I – I $_0$)** immediate successors that form a partition of it.

A node is a descendant of another if it derives from it directly or indirectly.

Every node is the top of a subtree of the search tree.

Any subtree contains all the descendants of its top, therefore, to delete a subtree, simply delete its top.

When exploring the search tree, the split-and-evaluate method uses a splitting procedure to successively partition the nodes.

If the passage from a node(0) to an immediate successor **D** is incompatible with the constraints of the problem to be solved, node **D** will be deleted. Otherwise, the split-and-evaluate method allows us to decide whether to explore or delete the subtree with vertex **D** . This tree is deleted if it has been judged useless to explore because it does not contain any solution better than the one available so far. The split-and-evaluate method is a typical example of the implicit enumeration approach, which can find an optimal solution by systematically examining sets of admissible solutions.

**Separation procedure**

This procedure describes the method used to partition a subset of possible solutions, the most common separation procedures are:

Backward branching :

The spots are affected one by one from the end.

Forward branching :

Tasks are assigned one by one from the beginning.

- At each stage, a task is chosen to be performed either at the beginning or at the end according to a certain heuristic method based on the problem data ( see)
- At each step, a task is chosen to be performed, either before or after another task. A heuristic may have been used to determine this pair of tasks.
- At each step, a task is chosen to be assigned either before or after another task. A heuristic can be used to determine this pair of tasks (see Hariri, [HAR 81]).

To minimize the criterion **f** of a particular scheduling problem, the split-and-evaluate method uses a procedure for calculating a lower bound called the default evaluation function.

**Procedure for calculating a lower bound:** •

This is a function defined on the set of nodes in the search tree, and denoted by **Lb** if **Y is** any node, we should be able to calculate

**Lb(Y)** and have: $\forall y \in Y \; Lb(Y) \leq f(y)$

Among the methods for calculating lower bounds for scheduling problems, we distinguish (for details,

Relaxation of problem constraints.

Lagrangian relaxation of constraints

Space " Relaxation of Dynamic Programming -

Lens relaxation. -

**Lower bounded**

If no admissible solution is known, the upper bound **UB** is initialized until the first admissible solution is found.

For a node **Y** that has not yet been deleted, we calculate its lower bound

**Lb (Y)**. If **Lb (Y) ≥UB** then we delete **Y**, otherwise **Y** is kept with a view to possibly exploring it later.

The search tree shrinks remarkably when the lower bounds are as large as possible to continue the search, the next node to explore is chosen according to a search strategy.

**Search strategy**

A search strategy allows to select in the tree the next node to explore, let **Y be** this searched node. By convention the first node to be explored is the root node **S.** Generally the node **Y** can be chosen as being:

The node with the smallest lower bound. -1

The most recently created node. -2

The node with the smallest lower bound among -3 the newly created nodes

Strategies (2) and (3) are depth-first strategy, while strategy (1) is breadth-first strategy.

The depth-first strategy requires little (very little) memory space, but a considerable amount of computation. On the other hand, the breadth-first strategy requires less computation because it chooses the branch to explore more intelligently and therefore finds an optimal solution faster. Unfortunately, this performance requires larger memory space requirements.

Regardless of the search strategy used, the separation and evaluation method will be completed once the search tree is fully explored and the remaining trial schedule will be optimal.

**Improved efficacity**

There are several ways to improve the efficiency of the separation and evaluation method, let us mention a few:

**Choice of a search strategy:** the strategy used is an important determinant of the time needed to solve a problem, generally breadth-first finds an optimal ordering faster than a depth-first strategy [FRE 82]

**Choice of lower bounds:**

The quality of the lower bounds is also an important factor in the computational time required to solve a problem, the lower bound **Lb (Y)** of node **Y** is good if it is not too small than the smallest value of the optimization criterion at that node, the split-and-flip method generally finds an optimal scheduling after examining fewer nodes with good lower bounds than with bad ones. The lower bounds eliminate the nodes that are at the top of the tree, which considerably reduces the search. Because when a node located at the top of the tree is eliminated, several subsequent nodes are eliminated at the same time.

**Rule of dominance:**

If it can be shown that an optimal solution can always be generated without finding a particular node in the search tree, that node is said to be dominated and can be eliminated.

In summary, dominance rules are useful in reducing the.

Computer memory requirements -

Computation time (because they avoid - computations for the descendants of any eliminated node).

**Initialization of the upper bound (UB):**

Starting with a near-optimal schedule can lead to the elimination of many nodes in the early stages of the search. Heuristic methods are used to find a near-optimal schedule, if any, an admissible schedule.

**Approximate methods**

Approximate methods are a very interesting alternative for dealing with large optimization problems if optimality is not essential; indeed, these methods have been used for a long time by many practitioners.

For large NP-hard problems, it is probably even impossible to find an optimal solution within a human lifetime, so in the case where we cannot find an optimal schedule within a reasonable time, we should not abandon analysis but rather use our experience to find in polynomial time a feasible solution, the best possible one.

Approximate methods are based mainly on various heuristics, often specific to a type of problem. Metaheuristics constitute another important part of approximate methods and open very interesting avenues in the design of heuristics for combinatorial optimization.

**Neighborhood methods**

Neighborhood methods are based on the notion of neighborhood.

**Definition**

Let **X be** the set of admissible configurations of a problem, we call neighborhood any application $N : x \to 2^{x}$, we call neighborhood exploration mechanism any procedure which specifies how the search goes from a configuration: **S** $\in$**X** to a configuration **S** $'\in$**N(s)** .

We say that the configuration **S** is a local optimum (minimum) concerning the neighborhood **N** if:

$$\forall S' \in N(S) : f(S) \leq f(S')$$

A typical neighborhood method starts with an initial configuration and then performs an iterative process that consists of replacing the current configuration with one of its neighbors taking into account the cost function, this process stops and returns the best configuration found when the stopping condition is met, this condition can generally be a limit for the iteration member.

## Tabu Research

The Taboo method was introduced simultaneously by several people ( P. Hansen, F. Glover , B. Jaumard ) around 1986.

Tabu search because it is forbidden to resume recently visited solutions.

At each iteration, the least bad neighbor is chosen.

To avoid cycles, that is to say, the infinite repetition of a sequence of movements, the last L movements are considered forbidden, L being the size of the Tabu list.

At each iteration, the movement made is therefore the least bad non-Tabumovement.

## Definition of Tabu Research (TR)

Is a metaheuristic method used for solving - optimization problems, primarily intended to guide other methods to find better solutions from an initial solution obtained by one of the heuristics.
The method can be used to guide any process that employs a set of moves to move from one solution to another and provides an evaluation function to measure the quality of those moves.

## Principle of Tabu Research

The principle of Tabu search is based on a method of moving through the solution space, while constantly seeking to improve the current best solution and keeping in memory the list of previous moves thus guiding the search outside of previously traveled areas.

The basic idea is inspired by search techniques used in artificial intelligence. It consists of keeping track of the past path of the search process in one or more memories and using this information to guide its future progress.

Basic principle: continue the search for solutions even when a local optimum is encountered and this:

By allowing movements that do not improve the solution.

Using the principle of memory to avoid backtracking (cyclical movements).

RT is based on:

The use of flexible memory structures (short, medium, long term) allows the complete exploration of the evaluation criterion and the search history.

A control mechanism based on alternating between conditions that restrict (Taboo restriction) and release (aspiration criterion) the search process.

The incorporation of so-called research intensification and diversification strategies:

The intensification strategy using medium-term memory serves to strengthen the search in the region for the best solutions found recently.

The diversification strategy using long-term memory is used to guide searches in new regions.

In general, we will not keep all the movements (too expensive in memory), but we will only prevent access to certain solutions during a certain number of iterations.

The neighborhood V(s(i)) of a solution is defined by an elementary transformation allowing one to move from a solution s(i) of S to another nearby solution s(j) with a small modification of the structure of the solution s(i). This operation is called movement.

A technique for searching only a subset of the neighborhood; then consists of keeping in a list of candidates a collection of movements likely to be better (of higher quality).

The simplest way to define Taboos is to keep a list of T.

When exploring a neighborhood, we choose the best neighbor excluding points from the Taboo list.

**Tabu List**

The Taboo list represents short-term memory, it contains the attributes of the most frequently performed movements.

This list is maintained for the purpose of guiding research.

**Aspiration criteria**

This criterion is often used to remove Tabu restrictions from a high-quality movement.

It allows the override of certain forbidden cases.
Its main use is to override the prohibition of a move if it allows to obtain an element better than the solution found so far (consists of revoking the Taboo status of a move if the latter allows to reach a solution of higher quality than that of the best solution found so far).

**Intensification criterion**

Intensification involves returning to one of the best solutions found so far and then resuming the search for that solution.

The intensification strategy is materialized in the following algorithm by strengthening the search in the list of best moves.

**Diversification criterion**

It consists of generating a new solution, different from those already explored, with the aim of going in a new direction, to explore another region.

**Stopping criterion**

As a general rule, to interrupt the algorithm, we take or leave two criteria into account, first of all, we check at each iteration that the total iteration counter has not exceeded a maximum number since the beginning of the process.

We base ourselves on another threshold value, which corresponds to the maximum number of iterations that we allow between two consecutive modifications (improvements) of the (best) solution.

But instead of considering the number of iterations, we could also base ourselves on the total time, which should be less than the maximum value.

**The general algorithm of the Tabu search**

Below we present the general algorithm of Tabu search, with emphasis on short-term memory and aggressive exploration:

(1) *Generate* an initial solution S randomly

(2) S* ←S ; C* ←F(S) / S* is the best solution encountered, C* is its cost and F the objective function

(3) *Add* S to the taboo list; K ←0

(4) **Repeat until** an end criterion is met

Choose from the neighborhood of $S_K$, $V(S_K)$,

the move which minimizes and which does not

belong to the taboo list, best($S_K$)

$S_{K+1} \leftarrow best(S_K)$ (6)

**If** the taboo list is full **then**

*Replace* the last element of the taboo list

with $S_{K+1}$

**If no**

Add $S_{K+1}$ to the taboo list

**End if** (11)

**If** ($C(S_{K+1}) < C^*$) **then**

$S^* \leftarrow S_{K+1}$, $C^* \leftarrow C(S_{K+1})$

**End if** (14)

**End** of algorithm

**Advantages and disadvantages of the taboo method**

**Benefits**

Tabu search has the advantage of being easily configurable. There are in fact at most two parameters that can really influence the search. The first is to choose the size of the tabu list. This is generally determined empirically and varies with the problems, but it is a crucial factor: too small a size can cause the algorithm to loop on a local area of the search space, while too large a size will quickly saturate the available resources. The second is the choice of the stopping criterion. This can be difficult to determine to avoid prolonging the search too long or missing the desired global optimum.

The effectiveness of the Tabu method means that it is widely used in combinatorial optimization problems: it has been successfully tested on major classical problems (traveling salesman, workshop scheduling) and is frequently applied to problems of planning, routing, geological exploration, etc.

Possibility of compromise between computation time and solution quality.

Opportunity to integrate problem-specific knowledge.

**Disadvantages**

metaheuristic methods have one thing in common: the lack of any guarantee of results. In fact, in the majority of cases, they do not provide any feedback on the quality of the solutions obtained.

Difficulty predicting performance (quality and time).

The Tabu method requires increasingly heavy memory management as one tries to refine the process by implementing complex memorization strategies.

**Evolutionary algorithms**

These algorithms are based on the principle of natural evolutionary processes and the mechanisms of evolution of living species.

A typical evolutionary algorithm consists of three essential elements:

A population consisting of several individuals representing solutions to the given problem.

A mechanism for assessing the adaptation of each individual in the population to their external environment.

An evolution mechanism composed of operators allows the elimination of certain individuals and the production of new individuals from the selected individuals.

An evolutionary algorithm starts with an initial population, often randomly generated, and repeats an evolutionary cycle consisting of three sequential steps.

Measure the adaptation (quality) of each individual in the population through the evaluation mechanism.

Select a portion of the individuals. Produce new individuals or recombinations of selected individuals.

This process ends when the stopping condition is met, for example, when a maximum generation member is reached. According to the analogy of natural evolution, the quality of individuals in the population should tend to improve as the processes progress.

An evolutionary algorithm includes: a set of operators such as selection, mutation, and possibly crossover. The objective of selection is to choose individuals who will be able to follow and reproduce to transmit their characteristics to the next generation. It is generally based on the conservation of the best-adapted individuals and the elimination of the least-adapted.

Crossbreeding seeks to combine the characteristics of parent individuals to create child individuals in the future generation and mutation makes slight modifications to some individuals.

Evolutionary algorithms can be divided into two broad classes: genetic algorithms and ant colony algorithms. These methods differ in their way of presenting information and the way in which the population evolves from one generation to the next.

**Genetic algorithm**

This class of methods is based on an imitation of the adaptation phenomena of living beings. Genetic algorithms operate on an analogy with the reproduction of living beings.

Generally speaking, genetic algorithms use the same principle: a population of individuals (corresponding to solutions) evolved at the same time, as in natural evolution in biology. Each individual, its ability to adapt to the external environment is measured by fitness. Genetic algorithms are then based on three functionalities.

**Selection:** this allows us to favor individuals who have better fitness for us, the fitness will be the most save the value of the objective function of the solution associated with the individual.

**Crossing:** which combines two parent solutions to form one or two children
( offspring ) trying to keep the good characteristics of the parent solutions.

**Mutation:** This allows diversity to be added to the population by mutating certain characteristics (genes) of a solution.

The representation of solutions (coding) is a critical point in the success of a genetic algorithm. It must, of course, adapt as well as possible to the problem and the evaluation of a solution.

**Algorithm**

1: Initialization: generate an initial population **P** of solutions of size |**P**| = **n**

2: Repeat

using a '3: Select: choose two solutions **X** and **X** selection technique

4: Crossover: combine the two parent solutions **X** and form a child solution y **toX**

5: Mutation of **y** under conditions

to be replaced in '6: Choose an individual solution y the population

with y in the population '7: Replace **y**

Until the stopping criterion is satisfied.

**Ant colony algorithms**

This new metaheuristic mimics the behavior of ants looking for food. Each time an ant moves it leaves a trace of its passage (the pheromone) as the ant is rarely an isolated explorer with several of its congeners it explores the region in search of food facing the obstacle, the group of ants explores both sides of the obstacle and finds itself, then they return to the nest with food. The other ants who want to obtain food will also take the same path if they separate facing the obstacle, the ants will then preferentially take the path on which the pheromone will be the strongest but the pheromone being an odor it evaporates. If few ants take a trace, it is possible that this path is no longer valid after a while, it is the same if explorer ants take a longer path (to bypass the obstacle by example ). On the other hand, if the path is heavily used, each new ant that passes redeposits a little pheromone and thus strengthens the trace, giving this path a greater possibility of being used.

**Algorithm**

1: Initialization: Create an initial population of ants

2: Repeat

3: For each ant do

4: Construct a solution by a construction procedure using pheromone traces.

5: Update of pheromone traces based on the quality of the solutions found.

6: End for

7: Until the stopping criterion is satisfied

**Hybrid methods**

The essential idea of this hybridization is to fully exploit the power of neighborhood search and recombination methods of evolutionary algorithms on a population of solution, such an algorithm uses one or more neighborhood methods on the individuals of the population during a certain iteration member or until the discovery of a set of local optima and then invokes a recombination mechanism to create a new individual as for specialized genetic algorithms, the recombination must imperatively be adapted to the problem treated.

Hybrid algorithms are arguably among the most powerful methods, but the computational time required can become prohibitive due to the number of individuals in the population being manipulated.

**Descent method**

This is one of the simplest local search heuristics. It consists of searching in the neighborhood of the current solution for a lower-cost solution. It proceeds in this way until a local optimum is reached.

Starting from a solution found by a heuristic, for example, we can very easily implement descent methods. These methods are all based on a simple principle: start with an existing solution, look for a solution in the neighborhood, and accept this solution if it improves the current solution.

The algorithm presents the skeleton of a simple descent method. Starting from an initial solution **x**, we choose a solution **x** ′in the neighborhood **N(x)** of **x**. If this solution is better than **x**, **f(x ′)** then we accept this solution as the new solution **x** and we repeat the process until there are no more improving solutions in the neighborhood of **x**.

**Algorithm**

1: initialization: find an initial solution **x**

2: repeat

3: neighborhood search: find a solution $x' \in N(x)$

4: if $f(x') < f(x)$ then

5: $x' := x$

6: end if

7: up to $f(y) \geq f(x), \forall y \in N(x)$

The main advantage of these methods lies in their great simplicity and speed. However, they have two major obstacles that significantly limit their effectiveness:

Depending on the size and structure of the - neighborhood **N(x)** considered, the search for the best neighboring solution is a problem that can be as difficult as the initial problem;

A descent method is unable to progress beyond the - first local minimum encountered. However, combinatorial optimization problems typically involve many local optima for which the value of the objective function can be very far from the optimal value.

To address these shortcomings, the simplest solution is the random relaunch method, which consists of randomly generating a new starting configuration and restarting a descent. Another solution is to accept neighbors with the same performance as the current configuration. This approach allows the search to move across plateaus but is not sufficient to emerge from all local optima. Other more elaborate refinements are also possible, for example: the introduction of variable neighborhoods, and neighborhood reduction or enlargement techniques

**Simulated Annealing**

This class of optimization methods is inspired by the simulation methods of Métropolies (1950s) in statistical mechanics. The historical analogy is inspired by the annealing of metals in metallurgy: a metal cooled too quickly presents numerous microscopic defects, this is the equivalent of a local optimum for a combinatorial optimization problem. If it is cooled slowly, the atoms rearrange, the defects disappear, and the metal then has a highly ordered structure, equivalent to a global optimum.

The analogy with an optimization method is found by associating a solution with a metal state, its thermodynamic equilibrium is the value of the objective function of this solution. Moving from one state of the metal to another corresponds to moving from one solution to a neighboring solution.

To move to a neighboring solution, one of the two conditions must be met:

- Either the movement improves the quality of the previous solution, **i.e.** in minimization, the negative cost variation ( $\Delta C < 0$).
- Let the movement deteriorate the quality of the previous solution and the probability **p** of accepting such a movement is less than a value dependent on the current temperature **t** ($p < e^{-\Delta C/t}$).

The temperature cooling scheme is one of the most difficult parts to adjust in this case. These schemes are crucial for obtaining an efficient implementation. Too fast cooling would lead to a local optimum that could be of very poor quality. Too slow cooling would be very costly in terms of computation time. Adjusting the various parameters (initial temperature, number of iterations per temperature step, temperature decrease, etc.) can therefore be long and difficult. Without being exhaustive, we usually encounter three main classes of cooling schemes.

- Stepwise reduction: Each temperature is held equal for a number of iterations, and thus decreases in steps.
- Continuous reduction: the temperature is changed at each iteration.
- Polynomial reduction: temperature decreases at each iteration with occasional increases.

Algorithm 2 presents the main characteristics of simulated annealing.

**Algorithm**

1: initialization: find an initial solution **x** , set an initial temperature **t**

2: repeat

3: Neighborhood search: find a solution $\mathbf{x}' \in \mathbf{N}(\mathbf{x})$

4: determine $\Delta C = \mathbf{f(x') - f(x)}$

5: get $\mathbf{p} \sim \mathbf{U(0.1)}$

6: if $\Delta C < 0$ or $e^{-\Delta C/t} > \mathbf{p}$ then

7: $\mathbf{x}' := \mathbf{x}$

8: end if

9: reduce the temperature **t** according to a cooling scheme

10: until a stopping criterion is satisfied.

**The table below shows some differences between exact and approximate methods.**

| **Approximate method** | **Exact method** |
|---|---|
| Effectiveness for large problems | Efficiency for special cases of reasonably sized problems |
| Reasonable memory space | Considerable memory space |
| Very short resolution time compared to exact methods | Solving time is usually considerable for large problems |
| No guarantee of optimality | Optimality guarantee |
| Practically simple to implement and | Practically difficult to implement and |

| | |
|---|---|
| understand | understand. |

**Example of a combinatorial problem**

**The scheduling problem**

The scheduling problem consists of organizing the time taken to complete tasks, taking into account time constraints (deadlines, sequencing constraints) and constraints relating to the use and availability of the required resources. -

A schedule is a solution to the scheduling problem; it describes the execution of tasks and the allocation of resources over time and aims to satisfy one or more objectives. More precisely, we speak of scheduling when we reserve the term sequencing for the case where only the relative order of tasks is fixed independently of the execution dates. -

**Tasks**

A task **i** is an elementary work entity (operation or set of operations), located in time by a start date ri or end date $\mathbf{C}_i$, the completion of which requires a duration $p_i = \mathbf{C}_i - \mathbf{r}_i$ and which consumes resources (material, personnel, monetary, etc.). -

-

| | |
|---|---|
| **I**: All tasks. ▪ | |
| **n:** the number of tasks. ▪ | |

| |
|---|
| **i:** task index. ▪<br>**t** $_i$**:** start date of execution of task **i**. ▪<br>**C** $_i$**:** end date of execution of task **i** . ▪<br>$p_i$**:** execution time of task **i.** ▪<br>$w_i$**:** weight of the stain i ▪ |

**The constraints**

In most scheduling problems, the tasks to be executed are subject to constraints that must be satisfied when searching for an optimal solution. There are three types of constraints.

**Potential constraints:**

These are the time location constraints, for example, "task **i** must precede task **j** " or task **i** must be completed before a certain date. With this type of constraint, the problem is called a "central scheduling problem".

**Disjunctive constraints**

When two tasks cannot be performed at the same time, we say that there is a disjunctive constraint that both tasks must satisfy.

**Cumulative constraints**

They concern the evolution over time of the total volume of human or material resources devoted to the execution of tasks.

**Resources**

Resources are the means required for the execution of tasks, they are of two types.

- A resource is consumable if, after being allocated to one task, it is no longer available for other tasks. Similar to the case of a raw material or money.
- A resource is renewable if, after being allocated to a task, it becomes available again, upon completion of that task, for other tasks, such as a machine, a process, a printer, etc.

**The criteria**

The criterion of a scheduling problem is the economic function that one aims to optimize. It is also called in another sense (objective function).

In general, the criterion is considered as an application **F** from the set $\Sigma$of permutations of the tasks of the problem posed to $\mathbf{R}^+$which, at each permutation $(j_1, j_2, \ldots j_n)$ associates a positive real number $F(j_1, j_2, \ldots j_n)$ and which expresses, in addition to the efficient use of resources, the overall execution time and compliance with the greatest number of constraints, the objective function associated with a scheduling problem is defined by: $F\,obj = \underset{\sigma\in\Sigma}{OPT}\ F(\sigma)$

**Examples of criteria**

**Total duration $C_{max}$**

In scheduling problems where the aim is to minimize the total execution time of tasks, the criterion defined by $C_{max}(\sigma) = \max_{i\in I} \{C_i\}$

Such that: **I** the set of tasks.

**i** is a task and $C_i$ is the completion date of task **i**,

$C_i = t_i + p_i$

This is therefore the execution date of the last task executed.

The objective function is written:

$$Fobj = \min_{\sigma \in \Sigma} \{C_{max}(\sigma)\}$$

**Overall delay:**

In scheduling problems where the aim is to minimize the overall task execution delay, the criterion is defined by:

$$T_{max}(\sigma) = \max_{i \in I} \{T_i\}$$

Where: is a permutation of tasks, **i** is a task, and $T_i$ is the execution delay of task **i**

$T_i = (0, C_i - d_i)$ **.**

The objective function is written:

$$Fobj = \min_{\sigma \in \Sigma} \{T_{max}(\sigma)\}$$

**The sum of delays: $\Sigma T_i$**

The objective function is written:

$$Fobj = \min_{i \in I} \left\{\sum T_i\right\}$$

**Weighted sum of delays: $\Sigma w_i T_i$**

The objective function is written:

$$Fobj = \min_{i \in I} \left\{\sum w_i T_i\right\}$$

$w_i$ unit cost (per unit of time) of delay of task **i** if it is completed late

$(C_i > d_i)$

**- Number of overdue tasks:** $\sum U_i$

The objective function is written: $Fobj = \min_{i \in I} \{\sum U_i\}$

$$U_i = \begin{cases} 1 & \text{if task i is completed late } (C_i > d_i) \\ 0 & \end{cases}$$

**- Weighted number of late tasks :** $\sum w_i U_i$

$$Fobj = \min_{i \in I} \{\sum w_i U_i\}$$

$w_e$ The unit cost (per unit time) of delay of task **i** if it is completed late

**($C_i > d_i$)**

**- The average delay:**

$$\overline{T} = \sum T_i / n$$

Where "n" is the number of spots. $Fobj = \min_{i \in I} \{\sum T_i / n\}$

**- The sum of the costs of advances and delays:**

$$Fobj = \min_{i \in I} \{\sum ( \alpha_i.E_i + \beta_i T_i )\}$$

**$E_i$:** advance of task **i** ( Earliness )

**$T_i$:** delay of spot **i** ( Tardiness )

**$\alpha_i$: the** advance unit cost associated with task **i**.

**$\beta_i$:** unit delay cost associated with task **i**.

**Regular optimization criterion**

Usually, optimization criteria are expressed in terms of the set of end dates of a schedule, so that their general form is always:

Where S is a schedule **σ(i** $Z(\sigma) = R(C_{\sigma(1)}, .........., C_{\sigma(n)})$

) is the task that is assigned to the **i** $^{th}$ position in σand **$C_i$** is the end date of task **i** in σ.

Let be $\sigma'$a schedule, we denote $C_{i'}$ by the end date of task **i** 'in $\sigma'$. The criterion Z is regular it is only if the following implication is true, whatever the schedule $\sigma$and$\sigma'$

$$\left[C_{\sigma(1)} \leq C_{\sigma'(1)},.......C_{\sigma(i)} \leq C_{\sigma'(i)}.........et\ C_{\sigma(n)} < C_{\sigma'(n)}\right] \Rightarrow \left[Z(\sigma) \leq Z(\sigma')\right]$$

**General characteristics of the schedules**

**Admissible scheduling**

scheduling is said to be admissible if it respects all the constraints of the problem (deadlines, precedents, resource limitations).

**Active scheduling**

In an active schedule, no task can start early without postponing the start of another.

**Semi-active scheduling**

Is a schedule in which the operation begins sooner without altering the sequence. In a semi-active schedule, a task cannot be advanced without changing the sequence of the resource.

**Scheduling without delay**

In zero-delay scheduling, the execution of a task should not be delayed if it is waiting and the resource is available.

**Representation of scheduling problems**

A classification can be made according to:

The number of machines and their order of use to manufacture a product depends on the nature of the workshop. A workshop is defined by the number of machines it contains and by its type. The different possible types are as follows:

**A machine ( $\phi$):** each job consists of a single operation.

**Parallel machines:** they perform, a priori, all the same functions, depending on their execution speed, we distinguish:

**Identical machines (P):** the execution speed is the same for all machines $M_i$ and all jobs $J_i$.

**Uniform machines (Q):** each machine $M_i$ has its own constant execution speed, the execution speed is the same for all jobs $J_i$ of the same machine $M_j$

**Independent machines (R):** the execution speed is different for each machine $M_j$ and for each job $J_i$.

**Dedicated machines:** these are specialized in carrying out certain operations. In this category, each job consists of several operations. Depending on the mode of passage of the operations on the different machines, three specialized workshops are differentiated, namely:

**Flow Shop (F):** this is a special case of the shop floor scheduling problem for which the work path is unique: The **n** jobs use the **m** machines in the order **1, 2, ………, m** ( production line ).

Most of the literature is limited to the special case of the permutation flow shop. A permutation schedule is a schedule that maintains the same order of jobs on all machines.

**Job Shop (J):** The **n** jobs must be executed on the **m** machines, under the same assumptions as the Flow Shop, the only difference is that the operating sequences relating to the different jobs can be distinct and are specific to each job.

**Open Shop (O):** This is a less constrained workshop model than the Flow Shop because the order of execution of operations is not fixed

**Rating**

In order to identify in a synthetic and precise manner the type of scheduling problem addressed, we use a three-field notation $\alpha/ \beta/ \gamma$:

$\alpha = \alpha_1 \alpha_2$ : describe the characteristics of the machines:

$\alpha_1 \in \{b, P, , R, O, F, J\}$

**b :** scheduling on a single machine.

**P:** scheduling on several parallel and identical machines.

**q :** scheduling on multiple machines and uniform.

**A:** Scheduling on multiple parallel and different machines.

**O:** This is an issue (Open Shop).

**F:** This is a problem (Flow Shop).

**J:** This is a problem (Job Shop).

$\alpha_2{}^{E} \{b, K\}$

**b :** the number of machines is variable.

**K:** the number of machines is K (K positive integer).

**B:** the set of constraints.

**γ:** the criterion to optimize.

**Examples**

$1//\sum w_i C_i$ : Represents the problem of minimizing the weighted sum of the end dates of independent tasks on a single machine.

$F3|V_i|\sum T_i$ : Represents the problems of minimizing the sum of delays in a 3-machine Flow Shop with availability date constraints.

**Gantt chart**

The Gantt chart is certainly the oldest type of representation, the most common, and the simplest for graphically visualizing task execution and resource utilization over time.

Each task is symbolized by a rectangle in which the code and duration of the task are entered, starting by representing the tasks with anteriority, a task is entered in the diagram when all its anteriorities have already been entered there; this representation clearly indicates the development of the work.

**Scheduling in the different types**

A very common classification of workshops, from a scheduling perspective, is based on the different machine configurations. The most well-known models are single-machine, parallel-machine, single-path workshop, and multi-path workshop.

**Single machine**

In this case, all the tasks performed are done by a single machine. The tasks then consist of a single operation that requires the same machine. One of the interesting situations where we can encounter this type of configuration is the case where we are faced with a production system including a bottleneck machine that influences the entire process. The study can then be restricted to the study of this machine.

**Theory of the complexity of algorithms**

Workshop scheduling problems are combinatorial problems, extremely difficult and there is no universal method to solve all cases efficiently, many of them can take a considerable time to solve, the theory of the complexity of algorithms has given a precise meaning to the term efficient algorithm and difficult problem, let us cite some basic concepts of this theory.

**The complexity of an algorithm**

Evaluating the complexity of an algorithm consists of finding, as a function of the size of the data, the number of elementary operations required by this algorithm.

**A polynomial algorithm**

Is an algorithm whose number of elementary operations necessary to solve an example of size **n** is a polynomial function in **n**.

**Non-polynomial algorithm**

Is an algorithm whose number of operations is not given by a polynomial of **n**.

**An efficient algorithm**

An algorithm is said to be efficient if the number of operations required to solve a problem is given by a polynomial function of a parameter characterizing the size of the problem considered.

**TABLE**

## REFFRENCE

**[1]** Fred Glove and Said Hanafi, 1999, Tabou Search and Finite Convergence, in Discrete Applied Mathematics.

**[2]** Grigoriev, M. Holthuijsen and J. V. De Klundertm 2005, Basic Scheduling Problems withRaw Materials Constraints, Naval Research Logistics.A.

**[3]** P. Hansen and N. Mladenovic, 2001, Rechrche à voisinage variable, Les cahiers De GERARD.

**[4]** Jin-Kao Hao et al., 1999, Métaheuristique pour l'optimisation Combinatoire et l'affectation sous contraintes, Revue d'Intelligence Artificielle vol. N° 1999

**[5]** M. Haouari and T. Ladhari, 2003, A Branch – and – Bound – Based Local Search MethodfortheFlowShopProbem,JORS,54,10 76–1084.

FSC
www.fsc.org
MIX
Papier | Fördert
gute Waldnutzung
FSC® C083411

Kolibri 360 GmbH
Ferdinand-Jühlke-Straße 7
99095 Erfurt, Deutschland
produktsicherheit@kolibri360.de

Druck:
CPI Druckdienstleistungen GmbH
im Auftrag der
Kolibri 360 GmbH
Ferdinand-Jühlke-Str. 7
99095 Erfurt